UNEP

RADIATION
EFFECTS and SOURCES

What is radiation?
What does radiation do to us?
Where does radiation come from?

D1692216

United Nations Environment Programme

DISCLAIMER

This publication is largely based on the findings of the United Nations Scientific Committee on the Effects of Atomic Radiation, a subsidiary body of the United Nations General Assembly and for which the United Nations Environment Programme provides the secretariat. This publication does not necessarily represent the views of the Scientific Committee or of the United Nations Environment Programme.

The designations employed and the presentation of the material in this publication do not imply the expression of any opinion whatsoever on the part of the United Nations Environment Programme concerning the legal status of any country, territory, city or area or of its authorities, or concerning delimitation of its frontiers or boundaries.

This publication may be reproduced in whole or in part and in any form for educational or non-profit purposes without special permission from the copyright holder, provided acknowledgement of the source is made. The United Nations Environment Programme would appreciate receiving a copy of any publication that uses this publication as a source.

No use of this publication may be made for resale or for any other commercial purpose whatsoever without prior permission in writing from the United Nations Environment Programme.

The United Nations Environment Programme promotes environmentally sound practices globally and in its own activities. This publication was printed on recycled paper, 100 per cent chlorine free. UNEP's distribution policy aims to reduce its carbon footprint.

Cataloguing: Radiation: effects and sources, United Nations Environment Programme, 2016

ISBN: 978-92-807-3517-8

Job No.: DEW/1937/NA

Copyright © United Nations Environment Programme, 2016

Printed in Austria

UNEP

RADIATION
EFFECTS and SOURCES

What is radiation?
What does radiation do to us?
Where does radiation come from?

United Nations Environment Programme

ACKNOWLEDGEMENTS

This booklet is largely based on the findings of the United Nations Scientific Committee on the Effects of Atomic Radiation and on the United Nations Environment Programme publication *Radiation: doses, effects, risks,* initially edited in 1985 and 1991 by Geoffrey Lean.

Technical editing: Malcolm Crick and Ferid Shannoun

Text editing: Susan Cohen-Unger and Ayhan Evrensel

Graphics and layout: Alexandra Diesner-Kuepfer

Furthermore, the following persons provided valuable contribution and comments on this booklet:

Laura Anderson, John Cooper, Susan Cueto-Habersack, Emilie van Deventer, Gillian Hirth, David Kinley, Vladislav Klener, Kristine Leysen, Kateřina Navrátilová-Rovenská, Jaya Mohan, Wolfgang-Ulrich Müller, Maria Pérez, Shin Saigusa, Bertrand Thériault, Hiroshi Yasuda, and Anthony Wrixon.

FOREWORD

Hiroshima, Nagasaki, Three Mile Island, Chernobyl and Fukushima-Daiichi: these names have become associated with the public's fear of radiation, either from use of nuclear weapons or accidents at nuclear power plants. In fact, people are much more exposed daily to radiation from many other sources, including the atmosphere and the Earth as well as from applications used in medicine and industry.

In 1955, nuclear weapon tests raised public concerns about the effects of atomic radiation on air, water and food. In response, the United Nations General Assembly established the United Nations Scientific Committee on the Effects of Atomic Radiation (UNSCEAR) to collect and evaluate information on the levels and effects of radiation exposure. The Committee's first report laid the scientific grounds for negotiating the Partial Test Ban Treaty in 1963 that prohibited atmospheric nuclear weapon testing. Since then, it has continued to produce high-profile reports on radiation exposure, including from the accidents at the Chernobyl and Fukushima-Daiichi nuclear power plants. The Committee has consistently delivered work of great value both to the scientific community and policymakers.

While the scientific community has published information on radiation sources and effects, it has tended to be technical and perhaps difficult for the general public to understand—which has often confused, rather than informed, the public, meaning that the fear and confusion engendered decades ago prevails. This publication tackles the issue by detailing the most up-to-date scientific information from UNSCEAR— on the types of radiation, their sources and effects on humans and the environment—and making it accessible to the general reader.

Today, the UNSCEAR secretariat operates under the auspices of the United Nations Environment Programme (UNEP), which helps countries implement environmentally sound policies and practices. Helping the public understand radiation and how it affects life on this planet lies within the core mandate of UNEP.

I am very pleased to congratulate all those who have contributed to this publication, as well as all the members of the Committee and their delegations, who have worked so diligently for the past six decades on these critical issues.

Achim Steiner
UNEP Executive Director and
Under-Secretary-General of the United Nations

CONTENTS

INTRODUCTION   1

1. WHAT IS RADIATION?   3
1.1.   Some history   3

1.2.   Some basics   4
       Radioactive decay and half-lives   6
       Radiation units   7

1.3.   Penetration power of radiation   9

2. WHAT DOES RADIATION DO TO US?   11
2.1.   Effects on humans   13
       Early health effects   14
       Delayed health effects   15
       Effects on offspring   18

2.2.   Effects on animals and plants   22

2.3.   Relationship of radiation doses and effects   24

3. WHERE DOES RADIATION COME FROM?   27
3.1.   Natural sources   28
       Cosmic sources   28
       Terrestrial sources   29
       Sources in food and drink   32

3.2.   Artificial sources   32
       Medical applications   33
       Nuclear weapons   37
       Nuclear reactors   39
       Industrial and other applications   48

3.3.   Average radiation exposure to public and workers   54

INTRODUCTION

Before we begin, we need to distinguish between ionizing and non-ionizing radiation. *Ionizing radiation* has enough energy to liberate electrons from an atom, thereby leaving the atom charged, whereas *non-ionizing radiation*, such as radio waves, visible light or ultra-violet radiation, does not. This publication is about the effects of radiation exposure from both natural and artificial sources. However, the word *radiation,* throughout, refers only to ionizing radiation.

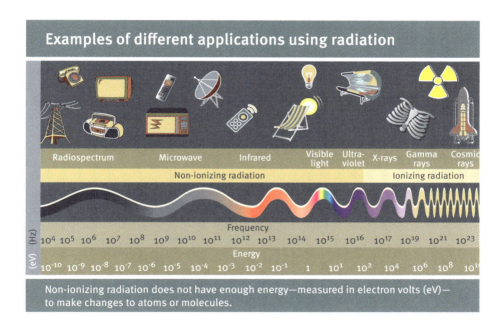

Examples of different applications using radiation

Radiospectrum	Microwave	Infrared	Visible light	Ultra-violet	X-rays	Gamma rays	Cosmic rays
Non-ionizing radiation				Ionizing radiation			

Frequency (Hz): 10^4 10^5 10^6 10^7 10^8 10^9 10^{10} 10^{11} 10^{12} 10^{13} 10^{14} 10^{15} 10^{16} 10^{17} 10^{19} 10^{21} 10^{23}

Energy (eV): 10^{-10} 10^{-9} 10^{-8} 10^{-7} 10^{-6} 10^{-5} 10^{-4} 10^{-3} 10^{-2} 10^{-1} 1 10^1 10^2 10^4 10^6 10^8 10^{10}

Non-ionizing radiation does not have enough energy—measured in electron volts (eV)—to make changes to atoms or molecules.

Today, we know more about the sources and effects of exposure to radiation than to almost any other hazardous agent, and the scientific community is constantly updating and analysing its knowledge. Most people are aware of the use of radiation in the nuclear power production of electricity or in medical applications. Yet, many other uses of nuclear technologies in industry, agriculture, construction, research and other areas are hardly known at all. To someone who is reading about the topic for the first time, it may come as a surprise that the sources of radiation causing the greatest exposure of the general public are not necessarily those that attract the most attention. In fact, the greatest exposure is caused by natural sources ever present in the environment, and the major contributor to exposure from artificial sources is the use of radiation in medicine worldwide. Moreover, everyday experience such as air travel and living in well-insulated homes in certain parts of the world can substantially increase exposure to radiation.

This publication is an attempt by the United Nations Environment Programme (UNEP) and the secretariat of the United Nations Scientific Committee on the Effects of Atomic Radiation (UNSCEAR) to help raise awareness and deepen understanding on the sources, levels and effects of exposure to ionizing radiation. Bringing together leading scientists from 27 Member States of the United Nations, UNSCEAR was set up by the United Nations General Assembly in 1955 to evaluate radiation exposures, effects and risks on a worldwide scale. However, it does not set, or even recommend, safety standards; rather it provides scientific information that enables national authorities and other bodies to do so. UNSCEAR's scientific evaluations over the past sixty years are the main source of information for this publication.

1. WHAT IS RADIATION?

To be able to talk about the levels, effects and risks of radiation exposure, we first need to address some basics of radiation science. Both radioactivity and the radiation it produces existed on Earth long before life emerged. In fact, they have been present in space since the beginning of the universe and radioactive material was part of the Earth at its very formation. But humanity first discovered this elemental, universal phenomenon only in the last years of the nineteenth century and we are still learning new ways of using it.

1.1. Some history

In 1895, **Wilhelm Conrad Roentgen**, a German physicist, discovered radiation—which he called X-rays—that could be used to look into the human body. This discovery heralded the medical uses of radiation, which have been expanding ever since. Roentgen was awarded the first Nobel Prize in physics in 1901 in recognition of the extraordinary services he had rendered to humanity. One year after Roentgen's discovery, **Henri Becquerel**, a French scientist, put some photographic plates away in a drawer with fragments of a mineral containing uranium. When he developed them, he found to his surprise that they had been affected by radiation. This phenomenon is called *radioactivity* and occurs when energy is released from an atom spontaneously and is measured today in units called becquerels (Bq) after Henri Becquerel. Soon afterwards, a young chemist, **Marie Skłodowska-Curie**, took the research further and was the first to coin the word radioactivity. In 1898, she and her husband **Pierre Curie** discovered that as uranium gave off radiation, it mysteriously turned into other elements, one of which they called polonium, after her homeland, and another they called radium, the "shining" element.

Wilhelm C. Roentgen (1845–1923)   Marie Curie (1867–1934)   Henri Becquerel (1852–1908)

Marie Curie shared the Nobel Prize in physics in 1903 with Pierre Curie and Henri Becquerel. She was the first woman to win the Nobel Prize a second time in 1911 for her discoveries in radiation chemistry.

1.2. Some basics

The scientists' quest was to understand the *atom* and, more particularly, its structure. We now know that atoms have a tiny, positively-charged nucleus surrounded by a cloud of negatively-charged *electrons*. The nucleus is only about one hundred thousandth of the size of the entire atom, but it is so dense that it accounts for almost the entire mass of the atom.

The nucleus is generally a cluster of particles, *protons* and *neutrons*, clinging tightly to each other. Protons have a positive electrical charge while neutrons have no charge. Chemical elements are determined by the number of protons in their atoms (e.g. boron has an atom with 5 protons and uranium has an atom with 92 protons). Elements with the same number of protons but a different number of neutrons are called *isotopes* (e.g. uranium-235 and uranium-238 differ in three neutrons in their nuclei). An atom as a whole is normally neither positively nor negatively charged because it has the same number of negatively-charged electrons as it has positively-charged protons.

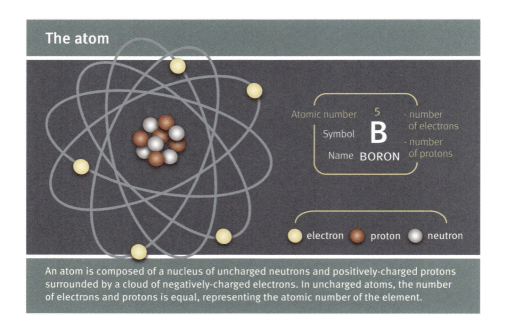

The atom

Atomic number 5 - number of electrons

Symbol **B**

Name **BORON** - number of protons

electron   proton   neutron

An atom is composed of a nucleus of uncharged neutrons and positively-charged protons surrounded by a cloud of negatively-charged electrons. In uncharged atoms, the number of electrons and protons is equal, representing the atomic number of the element.

Some atoms are naturally stable while others are unstable. Atoms with unstable nuclei—which spontaneously transform, releasing energy in the form of radiation—are known as *radionuclides*. This energy can interact with other atoms and ionize them. *Ionization* is the process by which atoms become positively or negatively charged by gaining or losing electrons. Ionizing radiation carries enough energy to knock electrons out of their orbit resulting in the creation of charged atoms called *ions*. The emission of two protons and two neutrons is referred to as *alpha decay* and the emission of electrons as *beta decay*. Frequently, the unstable nuclide will be so energized that the emission of particles is not sufficient to calm it down. It then gives off a vigorous burst of energy in the form of electromagnetic waves as photons called *gamma rays*.

X-rays are also electromagnetic radiation like gamma rays but with lower energy photons. An X-ray spectrum with different energies is produced in a vacuum tube made of glass when an electron beam, emitted by a *cathode*, is fired at target material called an *anode*. The X-ray spectrum depends on the anode material and the accelerating energy of the electron beam. Thus, X-rays can be generated artificially exactly when they are needed, which is very advantageous in industrial and medical applications.

X-ray tube

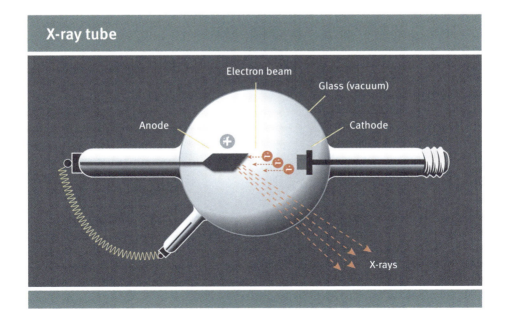

Electron beam

Glass (vacuum)

Anode

Cathode

X-rays

Radioactive decay and half-lives

While all radionuclides are unstable, some are more unstable than others. For example, the particles in the nucleus of a uranium-238 atom (with 92 protons and 146 neutrons) are only just able to cluster together. Eventually, a clump of two protons and two neutrons will break away and leave the atom as an alpha particle, turning the uranium-238 into thorium-234 (with 90 protons and 144 neutrons). But thorium-234 is also unstable, and transforms by a different process. By emitting high-energy electrons as beta particles and converting a neutron into a proton, it becomes protactinium-234, with 91 protons and 143 neutrons. This, in turn, is extremely unstable and soon becomes uranium-234, and so the atom goes on shedding particles and transforming itself until it finally ends up as lead-206, with 82 protons and 124 neutrons, which is stable. There are many such sequences of transformation, or *radioactive decay* as it is called.

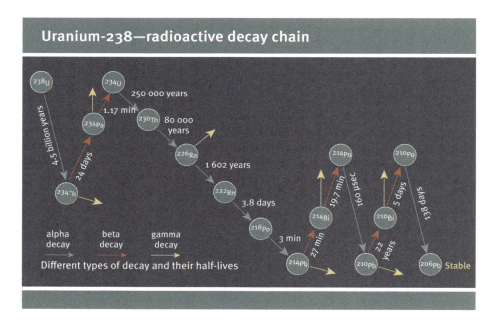

Uranium-238—radioactive decay chain

250 000 years
1.17 min
80 000 years
4.5 billion years
24 days
1 602 years
3.8 days
3 min
19.7 min
160 sec
5 days
138 days
27 min
22 years

alpha decay   beta decay   gamma decay

Different types of decay and their half-lives

238U · 234U · 234Pa · 230Th · 226Ra · 234Th · 222Rn · 218Po · 214Pb · 214Bi · 214Po · 210Pb · 210Bi · 210Po · 206Pb Stable

The period it takes half of any amount of an element to decay is known as its *half-life*. After one half-life, out of one million atoms on average 500 000 will decay into something else. During the next half-life about another 250 000 atoms will decay, and so on, until all have decayed. After 10 half-lives, only about a thousand remain of the original million (i.e. about 0.1 per cent). In the example given above, it would take a little over a minute for half of the atoms of protactinium-234 to decay to uranium-234. In contrast, for uranium-238 it would take four and a half billion years (4 500 000 000) for half of the atoms to decay to thorium-234. That said, only relatively few radionuclides occur naturally in the environment.

Radiation units

Today we know that the energy of radiation can damage living tissue, and the amount of energy deposited in living tissue is expressed in terms of a quantity called *dose*. The radiation dose may come from any radionuclide, or a number of radionuclides, whether they remain outside the body or irradiate it from inside, for example after being inhaled or ingested. Dose quantities are expressed in different ways depending on how much of the body and what parts of it are irradiated, whether one or many persons are exposed, and the length of the period of exposure (e.g. acute exposure).

Harold Gray (1905–1965)
Rolf Sievert (1896–1966)

The amount of radiation energy absorbed per kilogram of tissue is called the *absorbed dose* and is expressed in units called grays (Gy) named after the English physicist and pioneer in radiation biology, **Harold Gray**. But this does not give the full picture because the same dose from alpha particles can do much more damage than that from beta particles or gamma rays. To compare absorbed doses of different types of radiation, they need to be weighted for their potential to cause certain types of biological damage. This weighted dose is called the *equivalent dose*, which is evaluated in units called sieverts (Sv), named after the Swedish scientist **Rolf Sievert**. One sievert is 1 000 millisieverts, just as one litre is 1 000 millilitres or one meter is 1 000 millimetres.

Another consideration is that some parts of the body are more vulnerable than others. For example, a given equivalent dose of radiation is more likely to cause cancer in the lung than in the liver, and the reproductive organs are of particular concern because of the risk of hereditary effects. Thus, in order to compare doses when different tissues and organs are irradiated, the equivalent doses to different parts of the body are also weighted, and the result is called the *effective dose*, also expressed in sieverts (Sv). However, the effective dose is an indicator of the likelihood of cancer and genetic effects following lower doses and is not intended as a measure of severity of effects at higher doses.

This complex system of radiation quantities is necessary to bring them into a coherent structure, allowing radiation protection experts to record individual doses consistently and comparably, which is of major importance for people working with radiation and who are *occupationally exposed*.

Radiation quantities

Physical quantity	
Activity	The number of nuclear transformations of energy per unit of time. It is measured as decays per second and expressed in becquerels (Bq).
Absorbed dose	The amount of energy deposited by radiation in a unit mass of material, such as a tissue or organ. It is expressed in grays (Gy), which corresponds to joules per kilogram.
Calculated quantity	
Equivalent dose	The absorbed dose multiplied by a radiation factor (w_R) that takes into account the way different types of radiation cause biological harm in a tissue or organ. It is expressed in sieverts (Sv), which corresponds to joules per kilogram.
Effective dose	The equivalent dose multiplied by organ factors (w_T) that take into account the susceptibility to harm of different tissues and organs. It is expressed in sieverts (Sv), which corresponds to joules per kilogram.
Collective effective dose	Sum of all effective doses of a population or group of people exposed to radiation. It is expressed in man-sieverts (man Sv).

This, however, describes only doses to individuals. If we add up all the effective doses received by each individual in a population, the result is called the *collective effective dose* or simply *collective dose*, and this is expressed in man-sieverts (man Sv). For example, the annual collective dose to the world population is over 19 million man Sv corresponding to an annual average dose per person of 3 mSv.

1.3. Penetration power of radiation

In short, radiation may take the form of particles (including alpha, beta and neutron particles) or of electromagnetic waves (gamma rays and X-rays), all with different amounts of energy. The different emitting energies and particle types have different penetrating power—and so have different effects on living material. Since alpha particles are made up of two positively-charged protons and two neutrons, they carry the most charge of all radiation types. This increased charge means that they inter-act to a greater extent with surrounding atoms. This interaction rapidly reduces the energy of the particle and therefore reduces the penetrating power. Alpha particles can be stopped, for example, by a sheet of paper. Beta particles, made up of negatively-charged electrons, carry less charge and are therefore more penetrating than alpha particles. Beta particles can go through a centimetre or two of living tissue. Gamma rays and X-rays

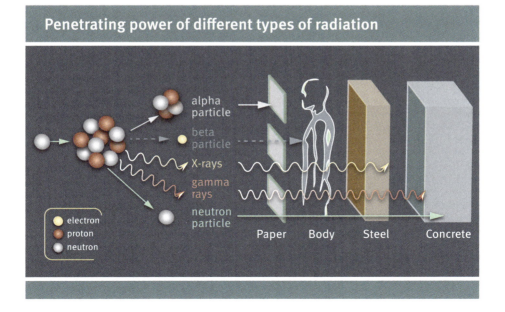

Penetrating power of different types of radiation

are extremely penetrating and will go through anything less dense than a thick slab of steel. Artificially produced neutrons are emitted from an unstable nucleus as a result of atomic fission or nuclear fusion. Neutrons can also occur naturally as a component of cosmic radiation. Because neutrons are electrically neutral particles, they have a very high penetrating power when interacting with material or tissue.

2. WHAT DOES RADIATION DO TO US?

Before going into more detail about the effects of radiation exposure, we should recall the pioneers in radiation science introduced earlier. Soon after *Henri Becquerel's* discovery, he himself experienced the most troublesome drawback of radiation—the effect it can have on living tissues. A vial of radium that he had put in his pocket damaged his skin.

Wilhelm Conrad Roentgen, who discovered X-rays in 1895, died of cancer of the intestine in 1923. *Marie Curie*, who was also exposed to radiation throughout her working life, died of a blood disease in 1934.

It is reported that by the end of the 1950s, at least 359 early radiation workers (mainly doctors and other scientists) had died from their exposure to radiation, unaware of the need for protection.

It is not surprising that those involved in applying radiation to patients were the first to develop recommendations for radiation protection of workers. By 1928, the International X-ray and Radium Protection Committee was set up during the second International Congress of Radiology in Stockholm and *Rolf Sievert* was elected as its first chair. After the Second World War—to take account of new uses of radiation outside medicine—it was restructured and renamed as the International Commission on Radiological Protection. Later on, between 1958 and 1960, Rolf Sievert was the fourth chair of UNSCEAR, at a time when there was particular concern about the genetic effects on humans from atomic weapon testing.

With the growing awareness of the risks associated with exposure to radiation, the twentieth century witnessed the development of intensive research on the effects of radiation on humans and the environment. The most important evaluation of population groups exposed to radiation is the study of approximately 86 500 survivors of the atomic bombings of Hiroshima and Nagasaki at the end of the Second World War in 1945 (henceforth referred to as *the survivors of the atomic bombings*). Further, reliable data on the subject comes from experience with irradiated patients, and with workers after accidental exposure (e.g. Chernobyl nuclear power plant accident), and from animal and cell experiments in laboratories.

Sources of information on radiation effects

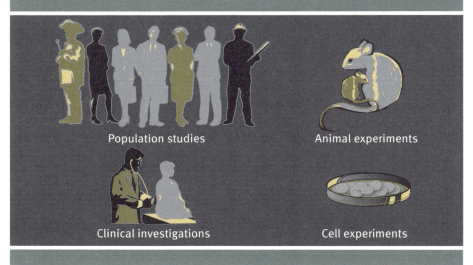

Population studies

Animal experiments

Clinical investigations

Cell experiments

UNSCEAR evaluates scientific information on effects of radiation exposure on humans and the environment and tries to work out, as reliably as possible, what effects can be associated with different levels of radiation exposure. As mentioned earlier, radiation exposure depends on the type of radiation, the time over which it is delivered and the amount of energy deposited in the material. For its evaluations, UNSCEAR currently uses the term *low dose* to mean levels below 100 mGy but greater than 10 mGy and the term *very low dose* for any levels below 10 mGy.

Dose bands used by UNSCEAR

High dose	More than ~1 Gy	Severe radiation accidents (e.g. firemen at the Chernobyl accident)
Moderate dose	~100 mGy to ~1 Gy	Recovery operation workers after the Chernobyl accident
Low dose	~10 mGy to ~100 mGy	Multiple computer tomography (CT) scans
Very low dose	Less than ~10 mGy	Conventional radiography (i.e. without CT)

2.1. Effects on humans

Since the discovery of radiation, more than a century of radiation research has yielded extensive information on the biological mechanisms by which radiation can affect health. It is known that radiation can produce effects at the level of cells, causing their death or modification usually because of direct damage to deoxyribonucleic acid (DNA) strands in a chromosome. If the number of damaged or killed cells is large enough, it may result in organ dysfunction and even death. Also, other damage to DNA may occur that does not kill the cell. Such damage is usually repaired completely but if not, the resulting modification— known as *cell mutation*—will be reflected in subsequent cell divisions and may ultimately lead to cancer. If the cells modified are those transmitting hereditary information to descendants, genetic disorders may arise. Information on biological mechanisms and on heritable effects is often gained from laboratory experiments.

Radiation damage to DNA strand

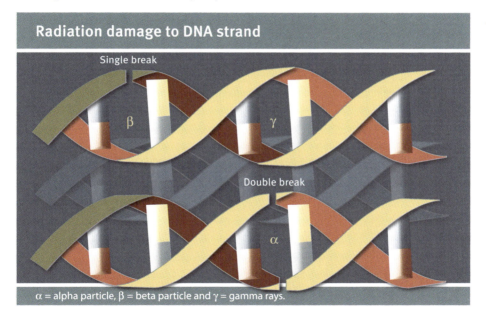

α = alpha particle, β = beta particle and γ = gamma rays.

On the basis of the observation of their occurrence, health effects following radiation exposure are defined here as either early or delayed health effects. Generally, early health effects are evident through diagnosis of clinical syndromes in individuals, and delayed health effects— such as cancer—through *epidemiological studies* by observation of increased occurrence of a pathology in a population. Further, special attention is paid here to effects on children and on embryos/fetuses, and to heritable effects.

Early health effects

Early health effects are caused by extensive cell death/damage. Examples are skin burns, loss of hair and impairment of fertility. These health effects are characterized by a relatively high threshold that must be exceeded over a short period before the effect occurs. The severity of the effect increases with increasing dose after the threshold has been exceeded.

Generally, acute doses higher than 50 Gy damage the central nervous system so badly that death occurs within a few days. Even at doses lower than 8 Gy, people show symptoms of radiation sickness also known as *acute radiation syndrome*, which could include nausea, vomiting, diarrhoea, intestinal cramps, salivation, dehydration, fatigue, apathy, listlessness, sweating, fever, headache and low blood pressure. The term acute refers to medical problems that occur directly after exposure rather than ones that develop after a prolonged period. However, victims may survive at first only to die from gastrointestinal damage one to two weeks later. Lower doses may not inflict gastrointestinal injury but still cause death after a few months, mainly from damage to the red bone marrow. Still lower doses will delay the onset of sickness and produce less severe symptoms. About half of those who receive doses of 2 Gy suffer from vomiting after about three hours, but this is rare for doses below 1 Gy.

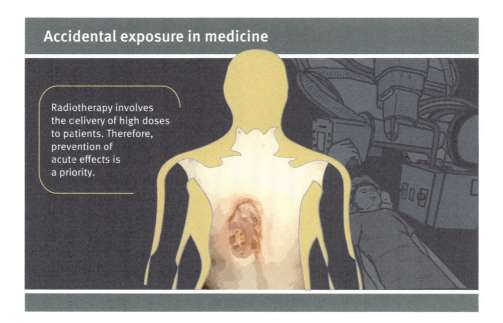

Accidental exposure in medicine

Radiotherapy involves the delivery of high doses to patients. Therefore, prevention of acute effects is a priority.

Fortunately, if the red bone marrow and the rest of the blood-forming system receive less than 1 Gy, they have a remarkable capacity for regeneration and can completely recover—although there will be a higher risk of developing leukaemia in later years. If only part of the body is irradiated, enough bone marrow will normally survive unimpaired to replace what has been damaged. Animal experiments suggest that even if only a tenth of the active bone marrow escapes irradiation, the chances of survival are nearly 100 per cent.

The fact that radiation can directly damage cell DNA is applied to deliberately kill malignant cells with radiation in cancer treatment known as *radiotherapy*. The total amount of radiation applied in radiotherapy varies depending on the type and stage of cancer being treated. Typical doses for solid tumour treatments range from 20 to 80 Gy to the tumour, which would endanger the patient if delivered as a single dose. Thus, in order to control the treatment, radiation doses are applied in repeated fractions of maximally 2 Gy. This fractionation allows cells of normal tissue to recover, while tumour cells are killed because they are generally less efficient at repairing after radiation exposure.

Delayed health effects

Delayed health effects occur a long time after exposure. In general, most delayed health effects are also stochastic effects, i.e. for which the probability of occurrence depends on the radiation dose received. These health effects are believed to be caused by modifications in the genetic material of a cell following radiation exposure. Examples of delayed effects are solid tumours and leukaemia occurring in exposed persons and genetic disorders occurring in the offspring of persons who were exposed to radiation. The frequency of occurrence—but not the severity—of these effects in a population appears to increase with larger doses.

Epidemiological studies are of great importance in understanding delayed health effects after radiation exposure. Such studies use statistical methods to compare the occurrence of a health effect (e.g. cancer) in an exposed population with that in an unexposed population. If a considerable increase is found in the exposed population, it may be that it is associated with the radiation exposure for the population as a whole.

The most important long-term evaluation of populations exposed to radiation is the epidemiological study of the survivors of the atomic bombings. This is the most comprehensive study ever conducted because of the large number of people, essentially representative of the general population, receiving a wide variety of doses spread fairly evenly over the body. Estimates of the doses received by this group are also relatively well known. So far, the study has revealed a few hundred more cancer cases than would be expected in this group if they had not been exposed to radiation. Because many of the survivors of the atomic bombings are still alive, studies are continuing in order to complete the evaluation.

Cancer

Cancer is responsible for about 20 per cent of all fatalities and is the most common cause of death in industrialized countries after cardio-vascular disease. About four out of ten persons in the general population are expected to develop cancer during their lifetime even in the absence of radiation exposure. In recent years, the most common cancers among men have been lung, prostate, colorectum, stomach, and liver cancer and among women they have been breast, colorectum, lung, cervix, and stomach cancer.

The development of a cancer is a complex process, consisting of a number of stages. An initiating phenomenon, most probably affecting a single cell, appears to start the process, but a series of other events seem to be necessary before the cell becomes malignant and the tumour develops. Cancer

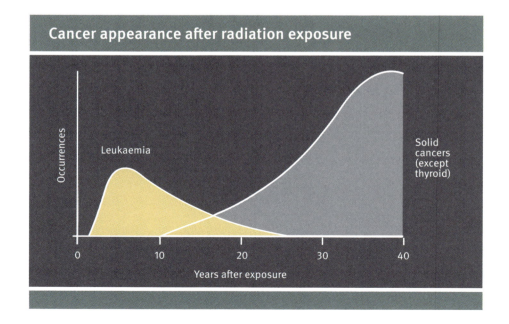

Cancer appearance after radiation exposure

Occurrences

Leukaemia

Solid cancers (except thyroid)

0   10   20   30   40

Years after exposure

becomes evident only long after the first damage is done, following a period of latency. The probability of cancer occurring following radiation exposure is a major concern and could be calculated for a group if it were exposed to a sufficiently high level of radiation to cause an increased occurrence of cancer that would overcome the statistical and other uncertainties. However, the real contribution of radiation as a cause of cancer remains unknown.

Leukaemia, thyroid cancer and bone cancer first appear within a few years of exposure to radiation, while most other cancers are not expressed until at least 10 years, often several decades, after exposure. However, no single type of cancer is uniquely caused by exposure to radiation so it is impossible to distinguish radiation-induced tumours from those arising from many other causes. Nevertheless, it is important to estimate the probability of getting cancer after certain doses of radiation in order to provide a sound scientific basis for setting exposure limits.

Studies of people who received medical treatment using radiation, people occupationally exposed, and—above all—the survivors of the atomic bombings build the foundation of the knowledge about the relationship between cancer and radiation exposure. These studies cover large samples of people who received exposure over many parts of the body and who were followed up over reasonably long periods. But some studies have major drawbacks, mainly a different age distribution from that of the normal population and the evidence that many of the these patients were already sick when irradiated and had already been receiving treatment for cancer.

More fundamentally still, almost all the data are based on the study of people whose tissues have received quite high doses of radiation, one gray or more, either as a single dose or over relatively short periods. There is little information on the effects of receiving low doses for a long time— just a few studies on the effects of the range of doses normally received by people working with radiation, and there is practically no direct information about the consequences of exposure to which the general public are routinely subjected. Studies would need to track a large number of people over a long period and eventually may still be too weak to observe increases in occurrence of cancer compared to the baseline cancer rates.

UNSCEAR conducted comprehensive reviews of the occurrence of cancer in populations exposed to radiation, and estimated that the additional chance of dying of cancer due to radiation exposure above 100 mSv was about 3 to 5 in a hundred per sievert.

Other health effects

High radiation doses to the heart increase the probability of cardiovascular diseases (e.g. heart attacks). Such exposure may happen during radiotherapy, although treatment techniques nowadays result in lower cardiac doses. However, there is no existing scientific evidence to conclude that exposure to low doses of radiation causes cardiovascular diseases.

UNSCEAR recognized that there was an increased occurrence of cataracts among Chernobyl emergency workers, possibly associated with high doses of radiation. Further, UNSCEAR has also studied the effects of radiation on the human immune system in survivors of the atomic bombings, in emergency workers at the Chernobyl nuclear power plant and in patients undergoing radiotherapy treatment. The effects of radiation on the immune system are assessed by estimating changes in cell numbers or by using a variety of functional analyses. High doses of radiation suppress the immune system mainly because of damage to lymphocytes. Their reduction is currently used as an early indicator to determine the radiation dose after acute exposure.

Effects on offspring

If radiation damage occurs in reproductive cells, the sperm or ovum, it can lead to heritable effects in descendants. Moreover, radiation directly can damage an embryo or fetus already developing within the womb. It is important to distinguish between radiation exposure of adults, children and embryos/fetuses. UNSCEAR has conducted comprehensive reviews of health effects, including heritable effects, in these groups.

Effects on children

Health effects in humans depend upon a number of physical factors. Because of their anatomical and physiological differences, the impacts of radiation exposure on children and on adults are different. Further, because children have smaller bodies and are less shielded by overlying tissues, the dose to their internal organs will be higher than that for adults for a given external exposure. Also, children are shorter than adults, so they may receive higher doses from radionuclides deposited on the ground.

Regarding internal exposure, because of the smaller size of children, and because their organs are, thus, closer together, radionuclides concentrated in one organ irradiate other organs more than would be the case for adults. There are also many other age-related factors involving metabolism and physiology that make a substantial difference in dose at different ages. Several radionuclides are of particular concern regarding internal exposure of children. Accidents involving releases of radioactive iodine-131 can be significant sources of exposure of the thyroid. For a given intake, the dose to the thyroid for infants is about nine times higher than that for adults. Studies of the Chernobyl nuclear power plant accident have confirmed the link between thyroid cancer and iodine-131, which mainly concentrates in this organ.

Epidemiological studies have shown that young people under 20 years of age appear to be about twice as likely as adults to develop leukaemia following the same radiation exposure. Further, children under 10 years are particularly susceptible; some other studies suggest that they are three to four times more likely to die of leukaemia than adults. Other studies have also shown that girls exposed at under 20 years of age are about twice as likely to develop breast cancer as adult women. Children are more likely than adults to develop cancer after radiation exposure, but it may not emerge until later in life when they reach an age at which the cancer normally becomes evident.

UNSCEAR has reviewed scientific material indicating that cancer occurrence in children is more variable than in adults and depends on tumour type, and on the child's age and sex. The term *radiosensitivity*

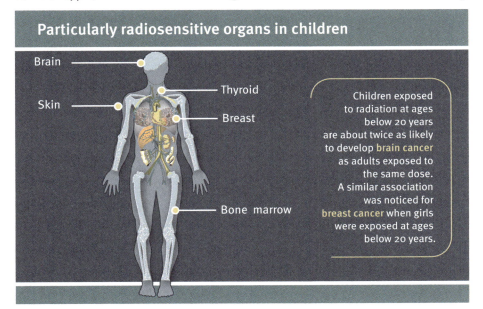

Particularly radiosensitive organs in children

Brain

Skin

Thyroid

Breast

Bone marrow

Children exposed to radiation at ages below 20 years are about twice as likely to develop brain cancer as adults exposed to the same dose. A similar association was noticed for breast cancer when girls were exposed at ages below 20 years.

with regard to cancer induction refers to the rate of tumours caused by irradiation. Studies on the differences in radiosensitivity between adults and children have found that children are more sensitive for the development of thyroid, brain, skin and breast cancer, and leukaemia.

Differences in early health effects on children following high doses (such as those received in radiotherapy) are complex and can be explained by the interaction of different tissues and biological mechanisms. Some effects are more evident for exposure in childhood than in adulthood (e.g. brain defects, cataracts and thyroid nodules); and there are a few effects for which children's tissues are more resistant (e.g. lungs and ovaries).

Effects on the unborn child

An embryo or fetus can be exposed through radioactive material transferred by the mother via food and drink (internal exposure) or directly through external exposure. Because a fetus is protected in the uterus, its radiation dose tends to be lower than the dose to its mother for most radiation exposure events. However, the embryo and fetus are particularly sensitive to radiation, and the health consequences of exposure can be severe, even at radiation doses lower than those that immediately affect the mother. Such consequences can include growth retardation, malformations, impaired brain function and cancer.

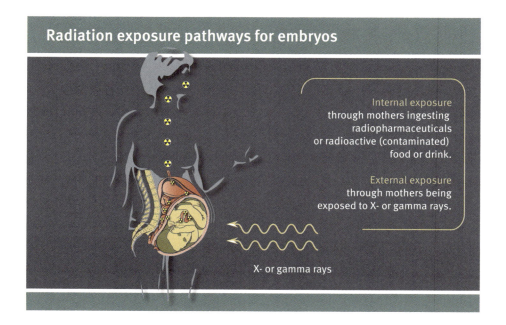

Radiation exposure pathways for embryos

Internal exposure through mothers ingesting radiopharmaceuticals or radioactive (contaminated) food or drink.

External exposure through mothers being exposed to X- or gamma rays.

X- or gamma rays

The development of mammals in the womb falls roughly into three stages. It is known that radiation might kill an embryo in the womb in the first stage, which lasts from conception to the point when it settles into the wall of the womb, and covers the first two weeks of pregnancy in humans. It is very hard to study what happens at this stage; however, information mainly from animal experiments confirms the fatal effect on the early embryo of radiation doses above certain thresholds.

During the next stage, lasting from the second to the eighth week in humans, the main danger is that radiation will lead to the growing organs becoming malformed and, perhaps, cause death at around the time of birth. Animal experiments have shown that organs (e.g. eyes, brain, skeleton) are particularly susceptible to malformation if irradiated just at the point when they are developing.

The greatest damage seems to occur in the central nervous system after the eighth week, when the third and last stage of pregnancy begins. Much progress has been made in understanding the effects of radiation exposure on brains of unborn children. As an example, 30 children of the survivors of the atomic bombings out of about 1 600 exposed before birth to a dose of 1 Gy had extreme intellectual disability.

There has been controversy about whether radiation exposure of embryos can cause cancer later in life. Animal experiments have failed to show any particular relationship. UNSCEAR has tried to estimate the overall risks to unborn children for a number of effects of irradiation—death, malformation, intellectual disability and cancer. In all, it reckons that no more than two out of every 1 000 live-born children who have been exposed to a dose of a hundredth of a gray in the womb, might be affected—compared with the 6 per cent who develop the same effects naturally.

Heritable effects

Radiation might modify cells transmitting hereditary information to descendants, which may cause genetic disorders. The study of such disorders is difficult because there is very little information on what genetic damage humans sustain through radiation exposure, partly because the full tally of heritable effects takes many generations to show, and partly because—as for cancer—these effects would be quite indistinguishable from those occurring from other causes.

Many of the severely affected embryos and fetuses do not survive. It has been estimated that about half of all miscarriages have an abnormal genetic constitution. Even if they do survive to birth, babies with genetic disorders are about five times more likely to die before their fifth birthday than are normal children.

Heritable effects fall into two main categories: chromosomal aberrations involving changes in the number or structure of chromosomes, and mutations of the genes themselves. They can appear in subsequent generations, but will not necessarily do so.

Studies of children whose parents were survivors of the atomic bombings have failed to find observable heritable effects. This does not mean that no damage has been sustained; just that moderate radiation exposure of even a relatively large population has no observable impact. However, experimental studies, in plants and animals exposed to high doses, have clearly demonstrated that radiation can induce heritable effects. Humans are unlikely to be an exception.

UNSCEAR has concentrated only on severe heritable effects and estimated that the total risk is about 0.3–0.5 per cent per gray—which is less than one-tenth of the possibility of occurrence of fatal cancer—to the first generation following radiation exposure.

2.2. Effects on animals and plants

Effects of radiation exposure on animals and plants are receiving more attention than previously. In past decades, the prevailing view was that if human life were adequately protected, both plants and animals would be similarly protected. UNSCEAR evaluated the effects of radiation exposure on plants and animals and found that a theoretical dose range of 1–10 Gy was unlikely to result in effects on animal and plant populations and that individual responses to radiation exposure varied (mammals are the most sensitive of all animals). Those effects that are likely to be significant at the population level concern fertility, mortality and the induction of mutations. *Reproductive changes*, such as in the numbers of offspring, are a more sensitive indicator of radiation effects than mortality.

Lethal doses represent doses at which 50 per cent of the exposed subjects would die. For plants exposed in a relatively short time (*acute*), these have been noted to range from less than 10 to about 1 000 Gy. In general, larger plants are more radiosensitive than smaller ones. Lethal doses range from 6 to 10 Gy for small mammals and are about 2.5 Gy for larger ones. Some insects, bacteria and viruses can tolerate doses of over 1 000 Gy.

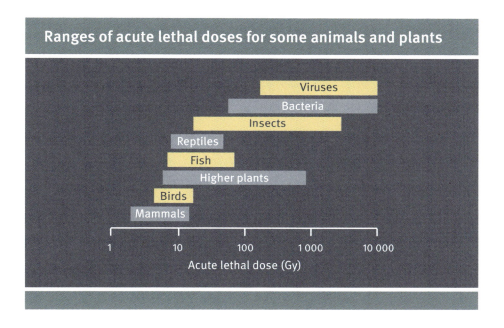

Ranges of acute lethal doses for some animals and plants

A main source of information has been the observations obtained from radiation exposure of animals and plants in the areas around the Chernobyl nuclear power plant. UNSCEAR evaluated pathways through which the environment was exposed, and developed new approaches for assessing the potential effects of such exposure.

Recently, UNSCEAR estimated doses and associated effects of radiation exposure for selected animals and plants after the Fukushima-Daiichi nuclear power station accident and concluded that the exposures were, in general, too low for acute effects to be observed. However, changes in *biomarkers*, which are indicators of a particular disease or physiological state of an organism—in particular for mammals—could not be ruled out, but their significance for the population integrity of those organisms was unclear.

It is important to note that protective and remedial action conducted to reduce radiation exposure to humans can have a significant wider impact. For example, it can affect environmental goods and services, resources used in agriculture, forestry, fisheries and tourism, and amenities used in spiritual, cultural and recreational activities.

2.3. Relationship of radiation doses and effects

When summarizing the relationship between radiation doses and health effects, UNSCEAR has stressed the importance of distinguishing between observations of existing health effects in exposed populations, and theoretical projections of possible future effects. For both situations, it is important to take into account any uncertainties and inaccuracies— whether in radiation measurements, statistical considerations or other factors.

Given the present state of knowledge, observed health effects can be confidently attributed to radiation exposure if early effects (e.g. skin burns) occur in individuals after high doses above 1 Gy. Such doses might arise in radiation accidents, such as those received by emergency workers during the Chernobyl nuclear power plant accident or by patients during accidents in radiotherapy.

It is possible, using epidemiological methods, to attribute an increased occurrence of delayed health effects (e.g. cancer) in a population exposed to moderate radiation doses if the observed increase is high enough to overcome any uncertainties. However, there are no biomarkers presently available to distinguish whether a cancer has been caused by radiation exposure or not.

Where the level of radiation exposure was low or very low—more typical of environmental and occupational radiation exposure—changes in the occurrence of delayed health effects have not been confirmed, given the statistical and other uncertainties. Nevertheless, such effects cannot be ruled out.

With regard to possible health effects in the future, there is an understanding of how to estimate the probability of the occurrence of these effects for high and moderate doses. However, at low and very low doses, it is necessary to make assumptions and use mathematical models to estimate the probability of any health effects resulting in values that are very uncertain. Consequently, for low and very low radiation doses, UNSCEAR

Relationship of radiation doses and health effects

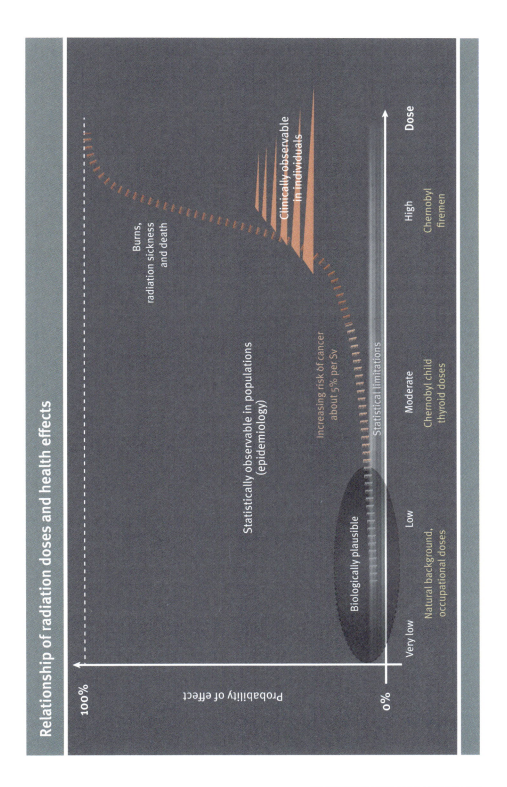

has chosen not to use such models in their assessments—following, for example, the Chernobyl and Fukushima-Daiichi accidents—to project numbers of health effects or deaths because of the unacceptable uncertainties in the predictions. Nevertheless, for public health comparisons or radiation protection purposes, it may be useful to make such calculations providing that the uncertainties are taken into account and the limitations are clearly explained.

3. WHERE DOES RADIATION COME FROM?

We are continuously exposed to radiation from many sources. All species on Earth have existed and evolved in environments where they have been exposed to radiation from the natural background. More recently, humans and other organisms have also been exposed to artificial sources developed over the past century or so. Over 80 per cent of our exposure is from natural sources and only 20 per cent is human-made from artificial sources—mainly from radiation applications used in medicine. Radiation exposure is categorized in this publication by its sources, with a focus on what the general public receive. For regulatory purposes (e.g. radiation protection) radiation exposure is addressed for different groups. Therefore, additional information is provided here on patients—who are exposed due to medical use of radiation—and on people exposed at workplaces.

Another way to categorize radiation exposure is how it irradiates us. Radioactive substances and radiation in the environment may irradiate our body from the outside—*externally*. Or we may inhale the substances in air, swallow them in food and water or absorb them through skin and wounds, and then they irradiate us from inside—*internally*. Considered globally, doses from internal and external exposure are about the same.

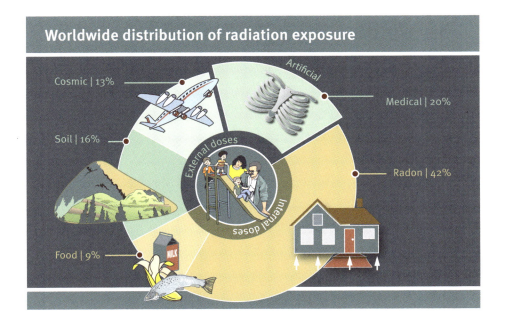

Worldwide distribution of radiation exposure

Cosmic | 13%

Artificial

Medical | 20%

Soil | 16%

External doses

Internal doses

Radon | 42%

Food | 9%

3.1. Natural sources

Since the creation of the Earth, its environment has been exposed to radiation both from outer space and from radioactive material in its crust and core. There is no way to avoid being exposed to these natural sources, which, in fact, cause most of the radiation exposure of the world's population. The global average effective dose per person is about 2.4 mSv and ranges from about 1 to more than 10 mSv depending on where people live. Buildings may trap a particular radioactive gas, called radon, or the building material itself may contain radionuclides that increase radiation exposure. Although the sources are natural, our exposure can be modified by choices we make, such as how and where we live or what we eat and drink.

Cosmic sources

Cosmic rays are a major natural source of external exposure to radiation. Most of these rays originate from deep in interstellar space; some are released from the sun during solar flares. They irradiate the Earth directly, and interact with the atmosphere, producing different types of radiation and radioactive material. They are the dominant radiation source in outer space. While the Earth's atmosphere and magnetic field considerably reduce cosmic radiation, some parts of the globe are more exposed than others. As cosmic radiation is deflected by the magnetic

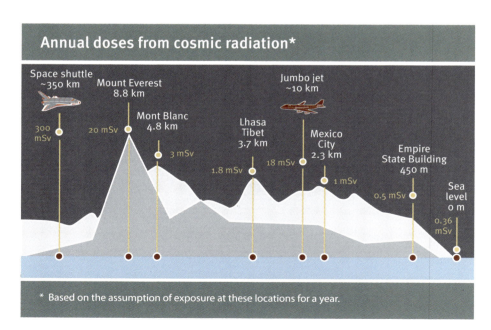

Annual doses from cosmic radiation*

Space shuttle ~350 km — 300 mSv
Mount Everest 8.8 km — 20 mSv
Mont Blanc 4.8 km — 3 mSv
Lhasa Tibet 3.7 km — 1.8 mSv
Jumbo jet ~10 km — 18 mSv
Mexico City 2.3 km — 1 mSv
Empire State Building 450 m — 0.5 mSv
Sea level 0 m — 0.36 mSv

* Based on the assumption of exposure at these locations for a year.

field to the North and South Poles, they receive more than the equatorial regions.

Moreover, the level of exposure increases with altitude because there is less air overhead to act as a shield. Thus, people living at sea level receive, on average, an effective dose of about 0.3 mSv annually from cosmic sources of radiation, or roughly 10–15 per cent of their total dose from natural sources. Those who live above 2 000 metres receive several times this dose. Airplane passengers might be exposed to even higher doses as the radiation exposure from cosmic sources depends not only on the altitude but also on the length of flights. For instance, at cruising altitudes, the average effective dose is 0.03–0.08 mSv for a 10 hour flight. In other words, a New York–Paris round trip flight would expose a person to about 0.05 mSv. This is approximately equal to the effective dose a patient would receive from a routine chest X-ray examination. Although the estimated effective doses received by individual passengers during a flight are low, the collective doses may be quite high because of the large number of passengers and flights worldwide.

EXPOSURE IN WORKPLACES

Doses from cosmic sources are particularly important for people who fly frequently such as pilots and cabin crew, who receive on average about 2–3 mSv annually. Doses have also been measured for a number of space missions. The reported doses for short space missions were in the range of 2–27 mSv, depending on solar activity. However, an astronaut on a four-month mission to the International Space Station which orbits the Earth at 350 km receives an effective dose of about 100 mSv.

Terrestrial sources

Soil

Everything in and on the Earth contains *primordial radionuclides*. These extremely long-lived radionuclides found in the ground—such as potassium-40, uranium-238 and thorium-232—together with the radionuclides into which they decay—such as radium-226 and radon-222—have been emitting radiation since before the Earth took its current shape. UNSCEAR calculates that every person worldwide receives, on average, an effective dose of about 0.48 mSv annually as external exposure from terrestrial sources.

External exposure varies considerably from one location to another. Studies in France, Germany, Italy, Japan and the United States, for example, suggest that about 95 per cent of their populations live in areas where the average annual dose outdoors varies from 0.3 to 0.6 mSv. However, in some places in these countries people can receive doses higher than 1 mSv annually. There are other places in the world where radiation exposure from terrestrial sources is higher still. For example, on the southwest coast of Kerala, India, a densely populated 55-kilometre long strip of land contains thorium-rich sands, where people receive, on average, 3.8 mSv annually. Other regions with high levels of natural terrestrial sources of radiation are known to exist in Brazil, China, the Islamic Republic of Iran, Madagascar, and Nigeria.

Radon gas

Radon-222 is a radionuclide in the form of a gas that normally emanates from the soil. It is produced from the decay series of uranium-238 present in the rocks and soil of the Earth. When inhaled, some of radon's short-lived decay products—mainly polonium-218 and -214—are retained in the lungs and irradiate cells in the respiratory tract with alpha particles. Radon is, hence, a primary cause of lung cancer in both smokers and non-smokers; however, smokers are far more vulnerable because of a strong interaction between smoking and radon exposure.

Radon is present in the atmosphere everywhere, and can seep directly into buildings through cellars and floors, where its *concentration*—the amount of activity in terms of decays per time in a volume of air—can build up. Mainly when homes are heated, warm air rises and escapes at the top of the house through windows or leakages, which creates low pressure in the ground floor and basement. This, in turn, causes active suction of radon from the subsoil through cracks and leakages (e.g. around service pipe entries) at the bottom of the house.

The worldwide-average concentration of indoor radon is about 50 Bq/m³. However, this average hides the great variability from place to place. In general, national average concentrations vary widely, ranging from less than 10 Bq/ m³ in Cyprus, Egypt and Cuba to more than 100 Bq/m³ in the Czech Republic, Finland and Luxembourg. In some countries such as Canada, Sweden and Switzerland there are houses with radon concentrations of between 1 000 and 10 000 Bq/m³. Nevertheless, the proportion of houses with such high-level concentrations is rare. Some of the factors that cause this variation are the underlying local geology, the permeability of the soil, the construction material and ventilation of buildings.

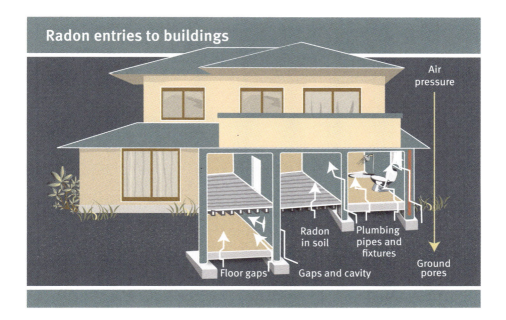

Radon entries to buildings

Air pressure

Radon in soil

Plumbing pipes and fixtures

Ground pores

Floor gaps

Gaps and cavity

In particular, ventilation, which depends on the climate, is a key factor. If buildings are well ventilated, such as in a tropical climate, the accumulation of radon is unlikely to be substantial. However, in temperate or cold climates, where places tend to be less ventilated, the concentrations of radon can build up considerably. Thus, the effect of restricted ventilation is important when designing energy-efficient buildings. Extensive measurement programmes have been conducted in many countries and have formed the basis for implementing measures to reduce indoor radon concentrations.

The level of radon in water is usually very low but some supplies—e.g. deep wells in Helsinki, Finland, and Hot Springs, Arkansas, United States—have very high concentrations. Radon in water can contribute to an increase of the concentration of radon in the air—particularly in the bathroom when showering. However, UNSCEAR concludes that the dose contribution from radon ingested in drinking water is small in comparison with its inhalation. UNSCEAR estimates that the average annual effective dose from radon is 1.3 mSv, representing about half of what the public receives from all natural sources.

For certain workplaces, inhaling radon gas dominates the radiation exposure of workers. Radon is the main source of radiation exposure in underground mines of all types. The annual average effective dose to a coal miner is about 2.4 mSv and for other miners about 3 mSv. In the nuclear industry, the annual average effective dose to a worker is about 1 mSv, mainly from radon exposure in uranium mining.

Sources in food and drink

Food and drink may contain primordial and some other radionuclides, mainly from natural sources. Radionuclides can be transferred to plants and then to animals from rocks and minerals present in the soil and water. Thus, the doses vary depending on the concentrations of radionuclides in food and water, and on local dietary habits.

For example, fish and shellfish have relatively high levels of lead-210 and polonium-210 and so people eating large amounts of seafood might receive somewhat higher doses than the general population do. Comparatively higher doses are also received by people in the arctic regions who consume large amounts of reindeer meat. Reindeer in the arctic contain relatively high concentrations of polonium-210, accumulated in the lichen they graze on. UNSCEAR estimates that the average effective dose from natural sources in food and drink is 0.3 mSv, due mainly to potassium-40 and to the uranium-238 and thorium-232 series radionuclides.

Radionuclides from artificial sources can be present in foodstuffs in addition to radionuclides from natural sources. However, the dose contribution from the authorized discharges of these radionuclides to the environment is usually very small.

3.2. Artificial sources

The uses of radiation have increased significantly over the past decades as scientists learned to use the energy of the atom for a wide variety of purposes, from military to medical applications (e.g. cancer treatment), and from electricity production to domestic applications (e.g. smoke detectors). These and other artificial sources add to the radiation dose from natural sources for both individuals and the global population.

Individual doses from artificial sources of radiation vary greatly. Most people receive a relatively small dose from such sources but a few receive many times the average. Artificial sources of radiation are generally well controlled by radiation protection measures.

Medical applications

The use of radiation in medicine to diagnose and treat certain diseases plays such an important role that it is now by far the main artificial source of exposure in the world. On average, it accounts for 98 per cent of the radiation exposure from all artificial sources and, after natural sources, is the second largest contributor to the population exposure worldwide, representing approximately 20 per cent of the total. Most of this exposure occurs in industrialized countries, where more resources for medical care are available and, therefore, radiology equipment is used much more extensively. In some countries, this has even resulted in an annual average effective dose from medical use that is similar to the one from natural sources.

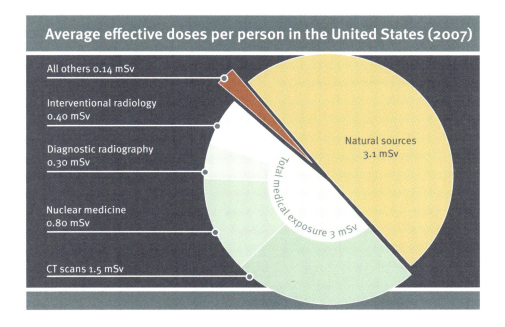

Average effective doses per person in the United States (2007)

All others 0.14 mSv

Interventional radiology 0.40 mSv

Diagnostic radiography 0.30 mSv

Nuclear medicine 0.80 mSv

CT scans 1.5 mSv

Natural sources 3.1 mSv

Total medical exposure 3 mSv

There are substantial and distinct differences between medical exposure and most other types of exposure. Medical exposure typically involves only a portion of the body, whereas other exposure often involves the whole body. Additionally, the distribution of patients' ages normally covers an older age range than that of the general population. Moreover, doses resulting from medical exposure should be compared with those from other sources very carefully, considering that patients receive a direct benefit from their exposure.

Increasing urbanization, together with a gradual improvement in living standards, inevitably means that more people can access health care. As a consequence, the population dose due to medical exposure continues to increase worldwide. UNSCEAR has been regularly collecting information on diagnostic and therapeutic procedures. According to its survey for the period 1997–2007, about 3.6 billion medical radiation procedures were performed annually worldwide, compared with 2.5 billion in the previous survey period covering 1991–1996, which is an increase of almost 50 per cent.

The main general categories of medical practice involving radiation are radiology (including interventional procedures), nuclear medicine and radiotherapy. Other uses not covered by UNSCEAR's regular evaluations include health screening programmes, and voluntary participation in medical, biomedical, diagnostic or therapeutic research programmes.

Diagnostic radiology is the analysis of images obtained using X-rays, such as in plain radiography (e.g. chest or dental X-rays), fluoroscopy (e.g. with barium meal or enema) and computer tomography (CT). Imaging modalities which use non-ionizing radiation, such as ultrasound or magnet resonance tomography, are not addressed by UNSCEAR. *Interventional radiology* uses minimally invasive image-guided procedures to diagnose and treat diseases (e.g. for guiding a catheter in a blood vessel).

Because of the wider use of CT and the significant dose per examination, the global average effective dose from diagnostic radiological procedures nearly doubled from 0.35 mSv in 1988 to 0.62 mSv in 2007. According to UNSCEAR's latest survey, CT scanning now accounts for 43 per cent of the total collective dose due to radiology. These numbers vary from region to region. About two thirds of all radiological procedures are received by the 25 per cent of the world's population living in industrialized countries. For the remaining 75 per cent of the world's population, the annual frequency of procedures has remained fairly constant, even for simple dental X-ray examinations.

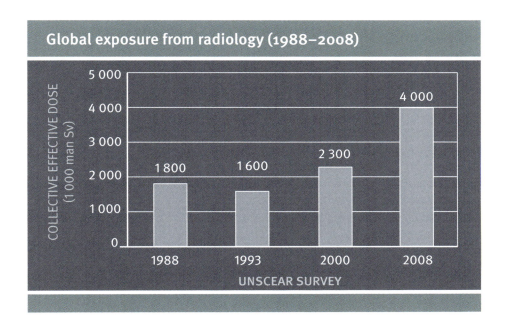

Global exposure from radiology (1988–2008)

COLLECTIVE EFFECTIVE DOSE (1 000 man Sv)

- 1988: 1 800
- 1993: 1 600
- 2000: 2 300
- 2008: 4 000

UNSCEAR SURVEY

Nuclear medicine is the introduction of *unsealed* (i.e. soluble and not encapsulated) radioactive substances into the body, mostly to obtain images that provide information on either structure or organ function and less commonly to treat certain diseases, such as hyperthyroidism and thyroid cancer. Generally, a radionuclide is modified to form a

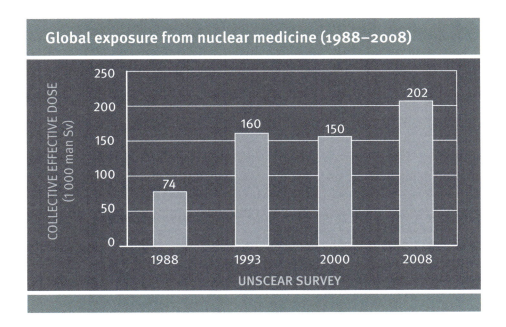

Global exposure from nuclear medicine (1988–2008)

COLLECTIVE EFFECTIVE DOSE (1 000 man Sv)

- 1988: 74
- 1993: 160
- 2000: 150
- 2008: 202

UNSCEAR SURVEY

radiopharmaceutical that is usually administered intravenously or orally. It then disperses in the body according to physical or chemical characteristics making a scan possible. Thus, the radiation emitted from the radionuclide within the body is analysed to produce diagnostic images or is used to treat diseases.

The number of diagnostic nuclear medicine procedures increased worldwide from about 24 million in 1988 to about 33 million in 2007. This resulted in a significant increase in the annual collective effective dose from 74 000 to 202 000 man Sv. Therapeutic applications in modern nuclear medicine are also increasing, reaching about 0.9 million patients each year worldwide. Again, the use of nuclear medicine is quite uneven, with 90 per cent of examinations occurring in industrialized countries.

Radiation therapy (also called *radiotherapy*) uses radiation to treat various diseases, usually cancer, but also benign tumours. External radiotherapy refers to patient treatment using a radiation source that is outside the patient's body and is called *teletherapy*. This uses a machine containing a highly radioactive source (usually cobalt-60) or a high-voltage machine that produces radiation (e.g. a linear accelerator). Treatment can also be performed by placing metallic or sealed radioactive sources, either temporarily or permanently, within the patient and this is called *brachytherapy*.

Worldwide, an estimated 5.1 million patients were treated annually with radiotherapy during the period 1997–2007, up from an estimated 4.3 million in 1988. About 4.7 million were treated by teletherapy and 0.4 million by brachytherapy. The 25 per cent of the population living in industrialized countries received 70 per cent of the radiotherapy treatment worldwide and 40 per cent of all brachytherapy procedures.

EXPOSURE IN WORKPLACES

Because the total number of medical radiological procedures has increased significantly in the past decades, so has the number of health workers involved, passing 7 million with an average annual effective dose of about 0.5 mSv per worker. In interventional radiology and nuclear medicine, medical staff might receive higher than the average dose.

Accidents in medical application

Some medical applications of radiation (e.g. radiotherapy, interventional radiology and nuclear medicine) involve the delivery of high doses to patients. When applied incorrectly, these can cause serious harm or even death. The people at risk include not only patients, but also physicians and other staff in the vicinity. Human error has been the most common cause of these accidents. Examples include giving a wrong dose because of treatment planning errors, failure to use equipment properly, and exposing the wrong organ or, occasionally, even the wrong patient.

While serious radiotherapy accidents are rare, over 100 have been catalogued. UNSCEAR has reviewed 29 reported accidents since 1967 that caused 45 deaths and 613 injuries. However, it is likely that some deaths and many injuries have not been reported.

Not only overexposure but underexposure might have serious consequences, when patients receive insufficient radiation dose to treat a life-threatening disease. Quality assurance programmes help to maintain high and consistent standards of practice in order to minimize the risk of such accidents.

Nuclear weapons

In 1945, during the final stage of the Second World War, two atomic bombs were dropped on Japanese cities—Hiroshima on 6 August and Nagasaki on 9 August. The explosions of the two bombs killed nearly 130 000 people. These events remain the only use of nuclear weapons for warfare in history. However, after 1945, many nuclear weapons were tested in the atmosphere, mostly in the northern hemisphere. The most active testing period was between 1952 and 1962. In all, over 500 tests were conducted, with a total yield of 430 megatons of trinitrotoluene (TNT) equivalent, the last in 1980. People throughout the world were exposed to radiation from the fallout from these tests. In response to concerns about the radiation exposure of humans and the environment, UNSCEAR was established in 1955.

The estimated annual average effective dose due to global fallout from atmospheric nuclear weapons testing was highest in 1963, at 0.11 mSv, and subsequently fell to its present level of about 0.005 mSv. This exposure will decline only very slowly in the future because most of it is now due to the long-lived radionuclide carbon-14.

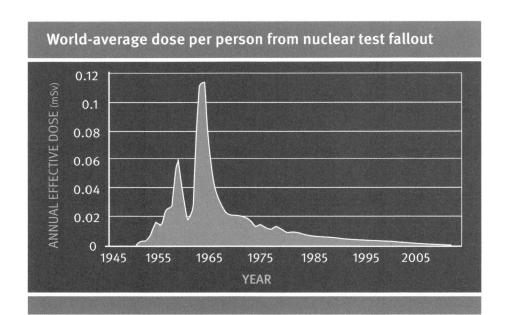

World-average dose per person from nuclear test fallout

As much as 50 per cent of the total fallout produced by surface tests was deposited locally within about 100 km of the test site. People living near test sites were thus exposed mainly to local fallout. However, because the tests were conducted in relatively remote areas, the local populations exposed were small and did not contribute significantly to the global collective dose. Nevertheless, people living downwind of the test sites received much higher doses than average.

UNSCEAR's first report in 1958 laid the scientific grounds on which the *Treaty Banning Nuclear Weapon Tests in the Atmosphere, in Outer Space and under Water* was negotiated. After the signature of this Partial Test Ban Treaty in 1963, about 50 tests were conducted underground annually until the 1990s; a few tests have also been conducted after that. Most of these tests had a much lower nuclear yield than the atmospheric tests, and any radioactive debris was usually contained unless gases were vented or leaked into the atmosphere. While the tests generated a very large quantity of radioactive residue, it is not expected to expose the public, because it is located deep underground and is essentially fused with the host rock.

There is concern regarding the reuse of nuclear test areas (e.g. for animal grazing or crop farming), because some are being reoccupied. The doses from radioactive residues at some sites, e.g. in localized areas at the Semipalatinsk test site in today's Kazakhstan, may be considerable, while in others, such as at the Mururoa and Fangataufa Atolls in French Polynesia, the doses would not contribute more than a fraction of the normal background exposure to a population eventually occupying the site. For other sites still, such as in the Marshall Islands and Maralinga, where the United States and the United Kingdom, respectively, conducted some of their tests, exposure of the population living there would depend on diet and lifestyle.

Nuclear reactors

When certain isotopes of uranium or plutonium are hit by neutrons, the nucleus splits into two smaller nuclei by a process called nuclear fission, releasing energy and two or more neutrons. The neutrons released may also hit other uranium or plutonium nuclei and cause them to split, releasing more neutrons, which in turn can split more nuclei. This is called a chain reaction. These isotopes are normally used as the fuel in nuclear reactors, where the chain reaction is controlled to stop it going too fast.

The energy released from fission in nuclear reactors can be used to produce electricity in nuclear power plants. However, there are also research reactors for testing nuclear fuel and various kinds of material, for investigations in nuclear physics and biology, and for the production of radionuclides to be used in medicine and industry. Although differences exist between the two types of reactors, both require industrial processes such as uranium mining and radioactive waste disposal, which can give rise to occupational and public exposure.

Nuclear power plants

The world's first commercial nuclear power station on an industrial scale, Calder Hall, was built in 1956 in the United Kingdom, and since then, the generation of electrical energy by nuclear power plants has grown considerably. Despite the increase in the decommissioning of older reactors, electrical energy production from nuclear sources continues to grow. By the end of 2010, around 440 power reactors were in operation in 29 countries, providing about 10 per cent of global electricity generation, and 240 research reactors were widespread worldwide in 56 countries.

Although the production of electricity by using nuclear power is often controversial, in normal operation it contributes very little to global radiation exposure. Moreover, the radiation exposure levels vary widely from one type of facility to another, between different locations and over time.

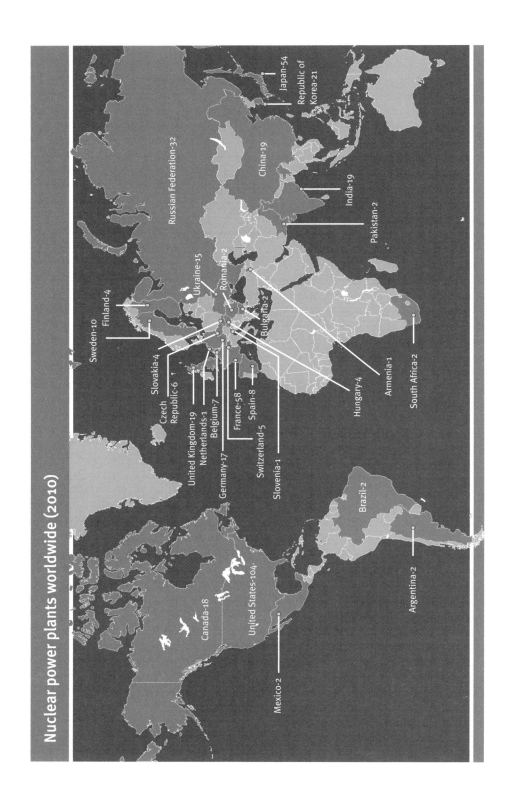

Nuclear power plants worldwide (2010)

Japan-54
Republic of Korea-21
Russian Federation-32
China-19
India-19
Pakistan-2
Ukraine-15
Romania-2
Bulgaria-2
Armenia-1
South Africa-2
Hungary-4
Finland-4
Sweden-10
Slovakia-4
Czech Republic-6
United Kingdom-19
Netherlands-1
Belgium-7
France-58
Spain-8
Germany-17
Switzerland-5
Slovenia-1
Brazil-2
Canada-18
United States-104
Argentina-2
Mexico-2

The overall exposure levels due to normal discharges from reactors have tended to decrease despite the increasing electrical output of plants. This is partly due to improvements in technology and partly due to stricter radiation protection measures. In general, discharges from nuclear facilities give rise to very low radiation doses. The annual collective dose to populations around nuclear power plants is estimated to be 75 man Sv. Thus, someone living in the vicinity of a power plant is exposed on average to an annual effective dose of about 0.0001 mSv.

The dominant component for radiation exposure from nuclear energy operations is mining. Uranium mining and milling produce substantial quantities of residues in the form of tailings, which contain elevated levels of natural radionuclides. By 2003, the total world production of uranium had reached about two million tonnes while the resultant tailings totalled over two billion tonnes. Current tailing piles are well maintained, but many old, abandoned sites exist and only a few have undergone remediation. UNSCEAR estimated the current annual collective dose to population groups around mine and mill sites and tailing piles at about 50–60 man Sv.

Spent fuel from reactors can be reprocessed to recover uranium and plutonium for reuse. Most spent fuel is currently retained in interim storage but about one third of that produced so far has been reprocessed. The annual collective dose due to reprocessing is estimated to be in the range of 20–30 man Sv.

Low-level and some intermediate-level waste is currently disposed of in near-surface facilities although, in the past, waste was sometimes dumped at sea. Both the high-level waste from reprocessing and the spent fuel (if not reprocessed) are stored but will eventually need to be disposed of. Proper disposal of the waste should not give rise to exposure of people even in the distant future.

EXPOSURE IN WORKPLACES

In the nuclear industry, the release of radon in underground uranium mines makes a substantial contribution to occupational exposure. The extraction and processing of radioactive ores that may contain high levels of radionuclides is a widespread activity. The average annual effective dose per worker in the nuclear industry has gradually declined since the 1970s, from 4.4 mSv to 1 mSv at present. This is mainly because of significant reduction in uranium mining coupled with more advanced mining techniques and ventilation.

Main processes in the nuclear industry

Conversion, enrichment and refinement prepares uranium for use as fuel.

Fuel fabrication produces fuel rods, generally from uranium in ceramic pellets, encased in metal tubes.

Research and power reactors, where the nuclei of uranium atoms split (fission) and release energy used to heat water.

Radioisotopes produced in reactors can be separated out for use in medicine and industry.

Milling extracts uranium from ore. The residues become tailings, containing long-lived radionuclides in low concentrations.

Reprocessing of uranium and plutonium from spent fuel can be recycled as fuel after conversion and enrichment.

Radioactive by-products make the fuel less efficient. After 12-24 months spent fuel is removed from the reactor.

Low- and intermediate-level waste is mostly disposed of in shallow burial sites on land.

Natural uranium is extracted mainly by open pit or underground mining.

High-level waste including spent fuel is currently held in interim storage pending final disposal in deep geological sites.

UF₆

Shallow and intermediate depth

Surface storage

Deep geological repository

Accidents at nuclear facilities

The exposure levels during the normal operation of civil facilities of the nuclear industry are very low. However, there have been some serious accidents, which received extensive public attention and whose consequences have been reviewed by UNSCEAR. Examples include the Vinca research facility in former Yugoslavia in 1958, the Three Mile Island nuclear power plant in the United States in 1979, and the fuel conversion facility at Tokai-Mura in Japan in 1999.

Severe radiation accidents in nuclear facilities between 1945 and 2007 resulted in 34 employee deaths or serious injuries, and seven caused off-site releases of radioactive material and detectable population exposure. There were other severe accidents in facilities related to nuclear weapon programmes. Excluding the 1986 Chernobyl and the 2011 Fukushima-Daiichi accidents—which are discussed below— 29 deaths and 68 cases of radiation-related injuries requiring medical care are known.

The most serious accident at a civilian installation before the one at Chernobyl was at the Three Mile Island nuclear power station on 28 March 1979. A series of events led to a partial meltdown of the reactor core. This accident released large amounts of fission products and radionuclides from the failed reactor core into the containment building, but relatively little into the environment and the resulting exposure of the public was very low.

Chernobyl nuclear power plant accident

The accident at the Chernobyl nuclear power plant on 26 April 1986 was not only the most severe in the history of civilian nuclear power, but also the most serious one in terms of exposure to radiation of the general population. The collective dose from the accident was many times greater than the combined collective dose from all other radiation accidents.

Two workers died from trauma in the immediate aftermath, and 134 suffered acute radiation syndrome, which proved fatal for 28 of them. Skin injuries and radiation-related cataracts were among the main problems for the survivors. Aside from the emergency workers, several hundred thousand people were subsequently involved in recovery operations. Apart from an apparent increase in occurrence of leukaemia and of cataracts among those who received high doses in 1986 and 1987, there is no consistent evidence to date of other radiation-related health effects in this group.

The accident caused the largest uncontrolled radioactive release into the environment ever recorded for any civilian operation; large quantities of radioactive substances were released into the atmosphere for about 10 days. The radioactive cloud created by the accident dispersed over the entire northern hemisphere and deposited substantial amounts of radioactive material over large areas of the former Soviet Union and other parts of Europe, contaminating land and water particularly in today's Belarus, the Russian Federation and Ukraine, and causing serious social and economic disruption to large segments of the population.

The contamination of fresh milk with the short-lived radionuclide iodine-131 (with a half-life of eight days) and the lack of prompt countermeasures led to very high doses to the thyroid, particularly of children, in parts of the former Soviet Union. Since the early 1990s, the occurrence of thyroid cancer among these persons exposed as children or adolescents in 1986 has increased in Belarus, Ukraine and four of the more affected regions of the Russian Federation. For the period 1991–2005, more than 6 000 cases had been reported; of these, 15 cases had proven fatal.

In the longer term, the general population was also exposed to radiation, both externally from radioactive deposits and internally from consuming contaminated foodstuffs, mainly by caesium-137 (with a half-life of 30 years). However, the resulting long-term radiation doses were relatively low, the average individual effective dose over the period 1986–2005 in contaminated areas of Belarus, the Russian Federation and Ukraine was 9 mSv. This is not likely to lead to substantial health effects in the general population. Still, the severe disruption caused by the accident has resulted in major social and economic impact and great distress for the affected populations.

UNSCEAR studied the radiological consequences of the accident in detail in several reports. The international community has made unprecedented efforts to assess the magnitude and characteristics of the accident's consequences in general and in different focus areas in order to improve understanding of the radiological and other consequences of the accident and assist in their mitigation.

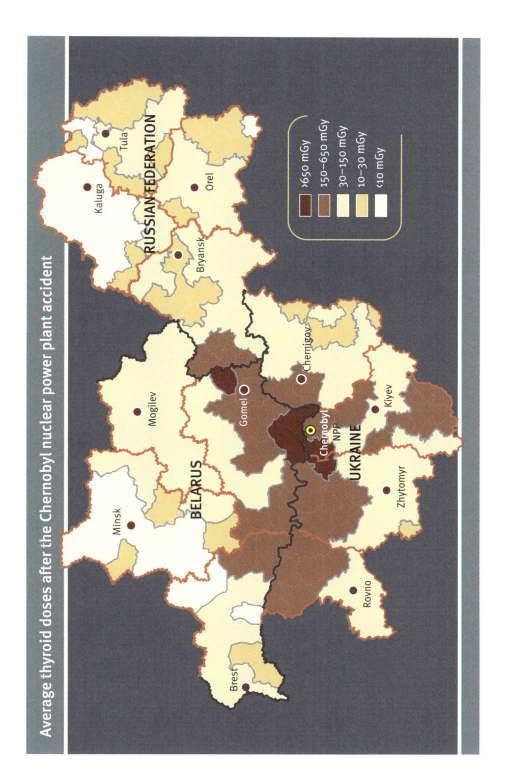

Average thyroid doses after the Chernobyl nuclear power plant accident

Legend:
- >650 mGy
- 150–650 mGy
- 30–150 mGy
- 10–30 mGy
- <10 mGy

RUSSIAN FEDERATION
- Tula
- Kaluga
- Orel
- Bryansk

BELARUS
- Mogilev
- Gomel
- Minsk
- Brest

UKRAINE
- Chernigov
- Chernobyl NPP
- Kiyev
- Zhytomyr
- Rovno

Essentially, studies since 1986 indicate that persons who were exposed as children to iodine-131 and the emergency and recovery operation workers who received high doses of radiation are at increased risk of radiation-induced effects. However, most area residents were exposed to low levels of radiation comparable to, or a few times higher than, the annual natural background radiation levels.

Fukushima-Daiichi nuclear power station accident

After the great east-Japan earthquake of magnitude 9.0 and tsunami on the east coast of northern Japan on 11 March 2011, the Fukushima-Daiichi nuclear power station was severely damaged and radioactive material was released to the environment. Approximately 85 000 residents within the 20-km area around the nuclear power station site and some nearby areas were evacuated as a precautionary measure between 11 and 15 March, while the residents living 20–30 km from the station were sheltered in their own homes. Later, in April 2011, the evacuation of another 10 000 people living further to the north-east of the station was recommended because of the elevated levels of radionuclides on the ground. These evacuations greatly reduced the levels of exposure that would have been received by those affected. The consumption of water and certain foodstuffs was temporarily restricted to limit the radiation exposure of the public. In managing the emergency situation at the nuclear power station, some operational staff and emergency response personnel were exposed.

UNSCEAR conducted an assessment of the radiation doses and associated effects on health and the environment. About 25 000 workers had been involved in mitigation and other activities at the Fukushima-Daiichi nuclear power station site during the first year and a half after the accident. The average effective dose to these workers for that time was about 12 mSv. However, 6 workers received cumulative total doses of over 250 mSv; the highest reported total dose was 680 mSv for one worker mainly received from internal exposure (about 90 per cent). Twelve workers were estimated to have received thyroid doses in the range of 2–12 Gy. No radiation-related deaths or acute diseases were observed among the workers exposed to radiation from the accident.

The average effective doses for adults in evacuated areas of the Fukushima Prefecture ranged from 1 mSv up to about 10 mSv in the first year after the accident. The effective doses for one-year-old infants were estimated to be about twice as high. For areas of the Fukushima Prefecture that were not evacuated and for neighbouring prefectures, the doses were lower.

Average thyroid doses after the Fukushima-Daiichi nuclear power station accident

Legend:
- 50–70 mGy
- 30–50 mGy
- 10–30 mGy
- <10 mGy
- Areas not assessed at district level
- Areas assessed separately

Fukushima-Daiichi NPP

YAMAGATA
MIYAGI
NIIGATA
FUKUSHIMA
IBARAKI
TOCHIGI
GUNMA

Estimates of the average doses to the thyroid, mainly from iodine-131, among those most exposed ranged up to 35 mGy for adults and up to 80 mGy for one-year-old infants. The annual dose to the thyroid, mainly from external natural radiation sources, is typically of the order of 1 mGy. UNSCEAR concluded a theoretical possibility that the risk of thyroid cancer among the group of children most exposed to radiation could increase. However, thyroid cancer is a rare disease among young children so that statistically no observable effects in this group are expected.

While comparisons are made with the Chernobyl disaster, the Fukushima-Daiichi nuclear accident was certainly different in terms of the type of reactor, the way the accident happened, the characteristics of the radionuclide releases and their dispersion, and the protective actions taken. In both cases, large amounts of iodine-131 and caesium-137—the two most significant radionuclides from the perspective of exposure after nuclear accidents—were released to the environment. The releases of iodine-131 and caesium-137 from the Fukushima-Daiichi accident compared to the Chernobyl one were about 10 and 20 per cent, respectively.

Industrial and other applications

Radiation sources are used in a broad spectrum of industrial applications. These include industrial irradiation used for sterilizing medical and pharmaceutical products, preserving foodstuffs or eradicating insect infestation; industrial radiography used for examining welded metal joints for defects; alpha or beta emitters used in luminizing compounds in gun sights and as low-level light sources for exit signs and map illuminators; radioactive sources or miniature X-ray machines used in well logging to measure geological characteristics in boreholes drilled for mineral, oil or gas exploration; radioactive sources used in devices to measure thickness, moisture, density and levels of material; and other sealed radioactive sources used in research.

Although widespread, the production of radionuclides for use in industrial and medical practices causes very low levels of exposure of the general public. However in the case of accidents, more localized areas can be contaminated and give rise to high levels of exposure.

EXPOSURE IN WORKPLACES

The number of workers involved in industrial uses of radiation was about one million in the early 2000s with an annual average effective dose per worker of 0.3 mSv.

Naturally occurring radioactive material

There are several types of facilities around the world that, while unrelated to the use of nuclear energy, may expose the public to radiation because of increased concentrations of *naturally occurring radioactive material* (NORM) in their industrial products, by-products and waste. The most important of such facilities involve mining and mineral processing.

Thickness measurement device using radiation

Activities related to the extraction and processing of ores can also lead to increased levels of NORM. These activities include uranium mining and milling; metal mining and smelting; phosphate production; coal mining and power generation from coal burning; oil and gas drilling; rare earth and titanium oxide industries; zirconium and ceramic industries; and applications using naturally-occurring radionuclides (typically isotopes of radium and thorium).

Coal, for example, contains traces of primordial radionuclides. Burning releases these radionuclides into the environment where they can expose people. This means that for each gigawatt year of electrical energy produced by the world's coal-fired power stations, the collective dose to the world population is estimated to increase by about 20 man Sv annually. In addition, fly ash (a residue generated in combustion) has been used in landfill and road construction, but using it for building construction results in radiation exposure from both direct irradiation and inhalation of radon. Furthermore, dumping fly ash may increase the radiation exposure levels around the dump site.

Geothermal energy generation is another source of radiation exposure of the general puplic. Underground reservoirs of steam and hot water are tapped to generate electricity or to heat buildings. Estimates of the emissions from the use of this technology in Italy and the United States suggests that it produces about 10 per cent of the collective dose per gigawatt year of electricity produced by coal-fired power stations. Geothermal energy currently makes a relatively small contribution to the world's energy production and thus to radiation exposure.

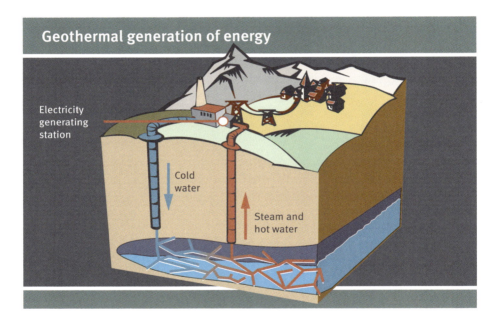

Geothermal generation of energy

Electricity generating station

Cold water

Steam and hot water

Various other human practices can expose people to NORM, such as the sludge from water treatment used in agriculture. However, exposure levels to the public are extremely low, of the order of less than a few thousandths of a millisievert annually.

A by-product of uranium enrichment is depleted uranium, which is less radioactive than natural uranium. Depleted uranium has been used for both civilian and military purposes for many years. Owing to its high density, it is used in radiation shielding or as counterweights in aircraft. Military use of depleted uranium, especially in armour-piercing munition, has raised concern about residual contamination. Except for a few specific scenarios, such as long-term handling, radiation exposure from depleted uranium is extremely low. In fact, chemical toxicity is its most hazardous property.

Consumer products

A number of products bought for everyday use contain low levels of radionuclides, deliberately added in order to make use of their chemical or radioactive properties. Historically, the most significant radionuclide for use in luminous consumer products was radium-226. This use ended several decades ago, with radium being replaced by promethium-147 and hydrogen-3 (tritium), which are less radiotoxic. Even so, for clocks or watches containing tritium compounds, some leakage of tritium may have occurred, because it is very mobile. However, tritium emits only very weak beta particles that cannot penetrate the skin so it exposes people only if the tritium entered the body.

Smoke detector function using radiation

Ionizing chamber

Screen

Metal plate

Smoke particles

Metal plate

BATTERY

Alpha particle

Radioactive source (e.g. americium)

Some modern smoke detectors consist of ionizing chambers with small foils of americium-241, which are emitters of alpha particles and produce a constant ion current. Ambient air is allowed to freely enter the detectors and if smoke enters the detector, it disrupts the current triggering an alarm.

The radioactivity of the americium source in a smoke detector is very low. It decays very slowly with a half-life of about 432 years. This means that a detector—at the end of 10 years' use—retains essentially all its original activity. As long as the americium source stays in the detector, exposure would be negligible. Although detectable with sensitive equipment, the exposure levels received from such products are extremely low. A person standing two metres away from the detector for eight hours a day is estimated to receive a dose of less than 0.0001 mSv a year.

Industrial accidents

Accidents involving industrial radioactive sources occur more often than those at nuclear power plants. Nevertheless, they do not normally receive as much attention even though they can cause extensive radiation exposure to both workers and members of the public.

Between 1945 and 2007, about 80 accidents have been reported at industrial facilities using radiation sources, accelerators and X-ray devices. Nine deaths were reported in these accidents, and 120 workers were injured. Acute radiation syndrome developed in some injured workers. The hands were a common site of injury, and often had to be amputated. UNSCEAR considers it probable that some accidents at industrial facilities involving deaths and injuries have not been reported.

The causes and effects of such accidents are many and varied. Just two examples are cited here. In 1978, in Louisiana, United States, an industrial radiographer working on a barge sustained a radiation injury to the left hand from a 3.7 TBq iridium-192 source, probably because of dosimeter malfunction. About three weeks later, his hand was red and swollen, and then skin blisters appeared, healing within 5–8 weeks. Six months later, however, the index finger had to be partially amputated. Then, in 1980, in Shanghai, China, because of improper safety measures, seven workers were exposed to radiation from a cobalt-60 source at an industrial facility. One worker, with a dose estimated at 12 Gy, died 25 days after exposure. A second, whose dose was estimated at 11 Gy, died 90 days after exposure. The other five workers received doses estimated to range from 2 to 5 Gy and recovered after medical treatment.

Orphan sources

Between 1966 and 2007, 31 accidents were credited to lost, stolen or abandoned radioactive sources, also known as *orphan sources*. These accidents are known to have resulted in the deaths of 42 members of the public, including children. In addition, acute radiation syndrome, serious local injuries, internal contamination or psychological problems necessitated medical care for hundreds of persons. Six accidents were associated with abandoned medical radiotherapy units.

Exactly how many orphan sources there are in the world is not known, but the numbers are thought to be in the thousands. The United States Nuclear Regulatory Commission reports that companies within the United States lost track of nearly 1 500 radioactive sources between 1996 and 2008, with more than half never recovered. A study by the European Union estimated that up to 70 sources are lost annually from regulatory control within its borders. Although the majority of these sources would not pose a significant radiological hazard, accidents are the major concern for orphan sources.

Worldwide estimates of serious radiation accidents*

Type of accident	1945–1965	1966–1986	1987–2007
Accidents at nuclear facilities	19	12	4
Industrial accidents	2	50	28
Orphan source accidents	3	15	16
Accidents in academia/research	2	16	4
Accidents in medicine	Unknown	18	14

* Based on accidents which have been reported officially or published. It is expected that the number of unreported accidents, especially in medicine, is much larger.

Sealed sources or their containers can be attractive to people who scavenge for the scrap metal trade because they appear to be made of valuable metal and may not display a radiation warning label. Cases where unsuspecting workers or even members of the public have tampered with sources have led to serious injury and in some cases death, as was the case in Goiânia, Brazil in 1987. An abandoned teletherapy device with a highly radioactive (50.9 TBq) caesium-137 source was stolen and the source capsule cracked open. Over the next two weeks, soluble caesium chloride powder was spread throughout a scrapyard and surrounding homes. Numerous people developed illnesses and skin lesions and 110 000 people had to be monitored for radioactive contamination, many of whom were internally contaminated with caesium-137. Because of this accident four people died, including one child.

3.3. Average radiation exposure to public and workers

Generally, public exposure to radiation from natural sources dominates the total exposure. UNSCEAR estimated the average annual effective dose to an individual at about 3 mSv. On average, the annual dose from natural sources is 2.4 mSv and two thirds of it comes from radioactive substances in the air we breathe, the food we eat and the water we drink. The main source of exposure from artificial sources is radiation used in medicine, with an individual average annual effective dose of 0.62 mSv. Medical radiological exposure varies by region, country and health-care system. UNSCEAR has estimated the average annual effective dose from medical applications of radiation in industrialized countries at 1.9 mSv and in non-industrialized countries at 0.32 mSv. However, these values might vary considerably (e.g. in the United States with 3 mSv or in Kenya with only 0.05 mSv).

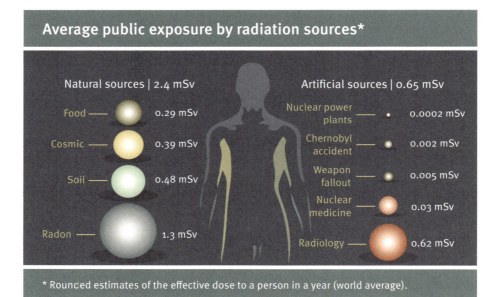

Average public exposure by radiation sources*

Natural sources | 2.4 mSv

Food — 0.29 mSv

Cosmic — 0.39 mSv

Soil — 0.48 mSv

Radon — 1.3 mSv

Artificial sources | 0.65 mSv

Nuclear power plants — 0.0002 mSv

Chernobyl accident — 0.002 mSv

Weapon fallout — 0.005 mSv

Nuclear medicine — 0.03 mSv

Radiology — 0.62 mSv

* Rounded estimates of the effective dose to a person in a year (world average).

Until the 1990s, attention concerning exposure of workers focused on artificial sources of radiation. Nowadays, however, it is realized that a very large number of workers are exposed to natural sources of radiation, mainly in the mining industry. For certain occupations in the mining sector, inhaling radon gas dominates radiation exposure at work. While the release of radon in underground uranium mines makes a substantial contribution to occupational exposure on the part of the nuclear industry, the annual average effective dose to a worker in the nuclear industry overall has decreased from 4.4 mSv in the 1970s to about 1 mSv today. However, the annual average effective dose to a coal miner is still about 2.4 mSv and for other miners about 3 mSv.

The current estimate of the total number of monitored workers is about 23 million worldwide, of whom about 10 million are exposed to artificial sources. Three out of four workers exposed to artificial sources work in the medical sector, with an annual effective dose per worker of 0.5 mSv. Evaluation of the trends of the average annual effective dose per worker shows an increase in exposure from natural sources mainly due to mining and a decrease in exposure from artificial sources mainly because of the successful implementation of radiation protection measures.

Trends in global radiological exposure of workers (mSv)*

Sources	1970s	1980s	1990s	2000s
Natural				
Aircrew	—	3.0	3.0	3.0
Coal mining	—	0.9	0.7	2.4
Other mining**	—	1.0	2.7	3.0
Miscellaneous	—	6.0	4.8	4.8
Total	—	1.7	1.8	2.9
Artificial				
Medical uses	0.8	0.6	0.3	0.5
Nuclear industry	4.4	3.7	1.8	1.0
Other industries	1.6	1.4	0.5	0.3
Miscellaneous	1.1	0.6	0.2	0.1
Total	1.7	1.4	0.6	0.5

  * Estimates of average effective dose per worker in a year.
** Uranium mining is included in nuclear industry.

UNSCEAR PUBLICATIONS

Since its inception, the United Nations Scientific Committee on the Effects of Atomic Radiation has issued over 25 major reports with over 100 scientific annexes, which are highly regarded as principal sources of authoritative evaluations examining radiation exposure from nuclear weapons tests and nuclear power production, medical usage of radiation, occupational radiation sources and natural sources. It also evaluates detailed studies on radiation-induced cancer and hereditary diseases, and assesses the radiological consequences of accidents on health and the environment. UNSCEAR reports and its scientific annexes are issued as United Nations sales publication (*unp.un.org*) and as free electronic downloads (*unscear.org*) to disseminate the findings for the benefit of UN Member States, the scientific community and the public.

Feedback and comments on this publication are highly appreciated to:

UNSCEAR secretariat
Vienna International Centre
P.O. Box 500
1400 Vienna, Austria
E-mail: *unscear@unscear.org*